TORNADOES:
CAUSES AND EFFECTS

By Kathleen Corrigan

12
STORY LIBRARY
MORE TO EXPLORE

ORION TOWNSHIP PUBLIC LIBRARY
825 Joslyn Rd Lake Orion MI 48362
(248) 693-3000
orionlibrary.org

www.12StoryLibrary.com

12-Story Library is an imprint of Bookstaves.

Developed and produced for 12-Story Library by Focus Strategic Communications Inc.

Library of Congress Cataloging-in-Publication Data
Names: Corrigan, Kathleen, author.
Title: Tornadoes : causes and effects / by Kathleen Corrigan.
Description: Mankato, Minnesota : 12-Story Library, [2022] | Series: Wild weather | Includes bibliographical references and index. | Audience: Ages 10–13 | Audience: Grades 4–6
Identifiers: LCCN 2020017489 (print) | LCCN 2020017490 (ebook) | ISBN 9781645821533 (library binding) | ISBN 9781645821915 (paperback) | ISBN 9781645822264 (pdf)
Subjects: LCSH: Tornadoes—Juvenile literature.
Classification: LCC QC955.2 .G33 2022 (print) | LCC QC955.2 (ebook) | DDC 551.55/3—dc23
LC record available at https://lccn.loc.gov/2020017489
LC ebook record available at https://lccn.loc.gov/2020017490

Photographs ©: Dan Ross/Shutterstock.com, cover, 1; Minerva Studio/Shutterstock.com, 4; Cammie Czuchnicki/Shutterstock.com, 5; skybaton/Shutterstock.com, 5; Rainer Lesniewski/Shutterstock.com, 6; Dana Mixer/ZUMApress.com/Alamy, 7; NASA, 7; A. T. Willet/Alamy, 8; Fer Gregory/Shutterstock.com, 9; imageBROKER/Alamy, 9; NOAA Photo Library/CC2.0, 10; U.S. National Weather Service, 11; Frode Jacobsen/Shutterstock.com, 11; NOAA National Severe Storms Laboratory/CC2.0, 12; desdemona72/Shutterstock.com, 13; TheAustinMan/CC4.0, 13; FLHC 3/Alamy, 14; NOAA Photo Library/CC2.0, 15; Jason Clark/Southcreek Global/ZUMApress.com/Alamy, 15; swa182/Shutterstock.com, 15; robertharding/Alamy, 16; NOAA Photo Library/CC2.0, 16; Big Joe/Shutterstock.com, 17; NOAA National Severe Storms Laboratory/CC2.0, 17; FEMA/Alamy, 18; Steve Skjold/Shutterstock.com, 19; Photoguru73/Shutterstock.com, 19; shae cardenas/Shutterstock.com, 19; Gene Blevins/ZUMApress.com/Alamy, 20; Roger Coulam/Alamy, 20; dcwcreations/Shutterstock.com, 21; Elena Larina/Shutterstock.com, 21; FEMA/Alamy, 22; Mary Hobbs/Alamy, 22; Noska Photo/Shutterstock.com, 23; Dmitry Kalinovsky/Shutterstock.com, 23; Cultura Creative/Alamy, 24; robertharding/Alamy, 24; World History Archive/Alamy, 25; PictureLux/The Hollywood Archive/Alamy, 25; Lennox Wright/Shutterstock.com, 26; VanoVasaio/Shutterstock.com, 26; Sajeev K A/Shutterstock.com, 27; Xolodan/Shutterstock.com, 27; Stuart Milliner/Shutterstock.com, 28; photka/Shutterstock.com, 28; dcwcreations/Shutterstock.com, 29; Tony-Gibson/Shutterstock.com, 29

About the Cover
A tornado funnel approaching a farm.

Access free, up-to-date content on this topic plus a full digital version of this book. Scan the QR code on page 31 or use your school's login at 12StoryLibrary.com.

Table of Contents

How a Tornado Is Born 4

Welcome to Tornado Alley 6

Ropes, Cones, and Pipes 8

Tornado Prediction Challenges 10

Tornado Detection Tools 12

History's Biggest Twisters 14

Warning the Public 16

Shelter in a Storm 18

A Trail of Destruction 20

After a Tornado 22

The Fascination of Tornadoes 24

Tornado Safety Myths 26

Staying Safe in a Tornado 28

Glossary 30

Read More 31

Index 32

About the Author 32

1

How a Tornado Is Born

Supercells develop into tornadoes.

Tornadoes are terrifying. They begin with supercell thunderstorms. When cold air, warm air, and moisture collide, supercells emerge. Cold air shoves warm air up. The moisture forms cloud towers. These towers produce heavy rain, hail, and lightning.

Meanwhile, wind churns through the atmosphere.

At different heights, it blows in different directions. The speeds are different, too. These wind currents stretch the bottom of the cloud towers. Under the supercell, cloud tubes twist across the sky. Rising warm air and sinking cold air tilt the tube. If the tube touches the ground, a tornado is born.

Tornado funnels are easy to identify.

Most tornadoes are 300 to 600 yards (275 to 550 m) wide. They may travel a short distance or many miles.

300
Wind speed in miles per hour (500 km/h) of fast tornadoes

- Scientists don't know the fastest wind speed of some tornadoes because the measuring tools are destroyed by the wind.
- Many tornadoes have wind speeds below 100 mph (160 kph).
- Sometimes tornadoes are called twisters.

JUST A DUST DEVIL

A dust devil is a small whirlwind. It gathers dust and dirt. It does not come from a cloud or thunderstorm. Instead, it rises from the hot ground. The heat twists with cool, sinking air.

Welcome to Tornado Alley

Tornado Alley stretches from Texas to Minnesota.

Tornadoes occur around the world. However, the US has far more tornadoes than any other country. Tornadoes have been reported in every state in the US. They are very rare in some states and common in others.

Many tornadoes happen in "Tornado Alley."

This is an area that gets very destructive tornadoes. It includes northern Texas, Oklahoma, Kansas, and Nebraska. Parts of South Dakota, Iowa, Minnesota and Colorado are also in Tornado Alley.

Tornadoes also impact the Southeast. "Dixie Alley" includes states on the Gulf Coast. Tennessee and Georgia are also included. Many people live on

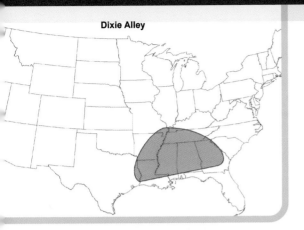

Dixie Alley

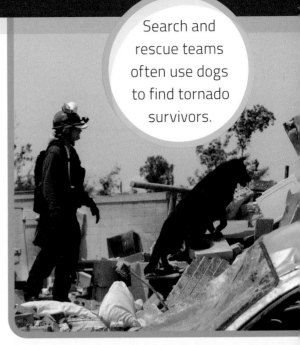

Search and rescue teams often use dogs to find tornado survivors.

the coast. As a result, Dixie Alley tornadoes are deadly. People may not see these tornadoes coming. The land is not flat and open like on the Plains.

THINK ABOUT IT

What causes air currents on Earth to spin?

1,000

Number of tornadoes in the US in a typical year

- Canada has the second most tornadoes, about 60 to 100 per year.
- Tornadoes in the Northern Hemisphere usually spin counterclockwise. Most tornadoes in the Southern Hemisphere spin clockwise.
- Texas gets the most tornadoes in a year. Alaska gets the fewest.

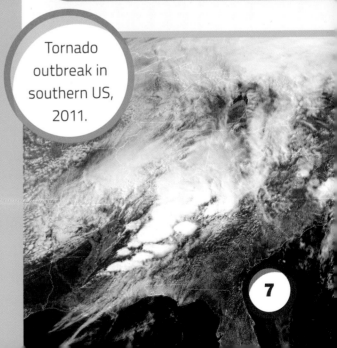

Tornado outbreak in southern US, 2011.

3
Ropes, Cones, and Pipes

Rope tornadoes are narrow but dangerous.

Tornadoes can happen anytime in the year. Most happen in the spring or summer.

They have different shapes and sizes. Rope tornadoes are twisty, narrow tornadoes. They are narrow but can still be dangerous. Funnel-shaped tornadoes may be cone tornadoes. They are narrow on the ground and wide at the top. Some tornadoes are like cylinders. They are known as stovepipe tornadoes.

8

Some tornadoes are colorless.

2.5

Width in miles (4 km) of a Nebraska tornado in 2004

- A 1925 tornado in Missouri, Illinois, and Indiana traveled over 200 miles (320 km).
- The center of a tornado might be calm like the eye of a hurricane.
- Many tornadoes occur between 3 and 9 pm.

Other tornadoes are like wedges. Major tornadoes are wedge tornadoes.

Tornadoes can be different colors. They may get color from the soil. They may pick up debris. Some look red or gray. Others are nearly colorless. Tornadoes can hide behind veils of dust, rain, and hail. A storm's darkness may hide them, too.

TORNADOES ON WATER

Tornadoes can also form over water. They are called *tornadic waterspouts*. Sometimes a waterspout will move onto land and cause damage.

Tornado Prediction Challenges

Meteorologists are scientists who study weather. They predict conditions that might lead to a tornado. They give tornado watches to let people know a tornado is possible. But they cannot tell for sure when or if a storm will produce a tornado. They cannot be sure what

Weather tracking stations monitor tornadoes.

Storm chasers track twisters.

900

Distance in miles (1,500 km) that some birds have flown to escape a storm

- So many storm chasers go to Oklahoma that they have traffic jams.
- In 2013, experienced storm chaser Tim Saramas and his team were killed studying a tornado.
- Some people pay to go on storm-chasing tours.

path a tornado will follow if it forms. Once a tornado forms, officials give warnings about where the storm may be headed. But scientists cannot know where a tornado will hit and cause damage.

Storm chasers study tornadoes up close. They try to find and follow tornadoes. Their job is exciting. It is also very dangerous. Some chasers are trained scientists. Others find extreme weather fascinating.

CAN ANIMALS PREDICT TORNADOES?

In 2014, tiny birds called golden-winged warblers left their nests in Dixie Alley. They flew away before a storm arrived. The storm brought 84 tornadoes. After the storm, they came back.

Tornado Detection Tools

Meteorologists use many tools to help them detect tornadoes forming. When supercells form, weather radar measures the amount of precipitation in the clouds. Anemometers track wind speeds. Scientists use Doppler radar to determine the speed and direction of swirling wind currents. When the supercell curls into a hook shape, tornadoes may form. Officials then issue a tornado *watch*. Once a tornado is spotted, they issue a *warning*.

Tornadoes are classified into different groups. Wind speeds and destruction set them apart. The Enhanced Fujita Scale (EF-Scale) rates tornadoes. Tornadoes can be EF0 to EF5. An EF0 storm causes minor damage. It may

Rating	Wind Speed	Damage
EF0	65-85mph	minor roof, branches
EF1	86-110	broken windows
EF2	111-135	roofs off, large trees
EF3	136-165	homes damaged
EF4	166-200	homes leveled
EF5	200+	incredible damage

TORNADO RATING
Enhanced Fujita Scale

148
Number of tornadoes that occurred on one day in 1974 in the US

- Usually there are about 15 minutes between a tornado warning and its arrival.
- Volunteers join SKYWARN and watch for tornado signs. Their information goes to the National Weather Service.
- In 2011, an SUV was thrown about a half mile (800 m) during an EF5 tornado in Smithville, Mississippi.

push over some trees and peel shingles off roofs. An EF3 can crush shopping malls. It can knock over trains and toss cars. EF5 storms cause incredible damage. They collapse tall buildings. They plow houses flat.

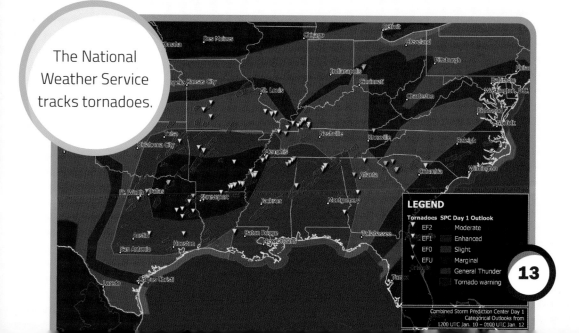

The National Weather Service tracks tornadoes.

LEGEND

Tornadoes	SPC Day 1 Outlook
EF2	Moderate
EF1	Enhanced
EF0	Slight
EFU	Marginal
	General Thunder
	Tornado warning

Combined Storm Prediction Center Day 1 Categorical Outlooks from 1200 UTC Jan. 10 – 0100 UTC Jan. 12

6

History's Biggest Twisters

FIRST PICTURES OF STORM DISASTER

HERALD EXAMINER

Telephone Main 5000   FRIDAY, MARCH 20, 1925.   C*   TWO PARTS   PRICE 3 CENTS

1,000 DEAD, 3,000 HURT LATEST TOLL OF TORNADO

In the Twirkling of an Eye, Murphysboro Was No More

A newspaper exaggerates the number of dead and injured people in the Tri-State Tornado

The worst recorded tornado in the world was in Bangladesh in 1989. Over 1,300 people died. Twelve thousand others were hurt when the mile-wide (1.6 km) tornado hit.

The Tri-State Tornado in 1925 was the deadliest tornado in the US.

It killed 695 people in Missouri, Illinois, and Indiana. The tornado lasted 3.5 hours. It traveled 219 miles (352 km).

The 1974 Super Outbreak affected 13 states in the US. There were 148 tornadoes. Seven were EF5. Over 20 were EF4 tornadoes. They caused

The 1974 tornados left destruction in their wake.

about a billion dollars in damages. Over 330 people were killed.

The 2011 Super Outbreak in the southeastern US was more destructive. In four days, 360 tornadoes churned. The supercell responsible swelled across 380 miles (610 km).

80,000
Number of people left homeless after a Bangladesh tornado in 1989

- The 1925 Tri-State Tornado smashed 15,000 homes.
- In 1974, a tornado tossed school buses on top of the high school in Xenia, Ohio.
- A 2011 tornado dug a trench 2 feet (.6 m) deep in Mississippi.

WILL CLIMATE CHANGE AFFECT TORNADOES?

Climate change may impact tornado patterns. Scientists are not sure yet. Climate change may cause more severe storms. These could create more tornadoes.

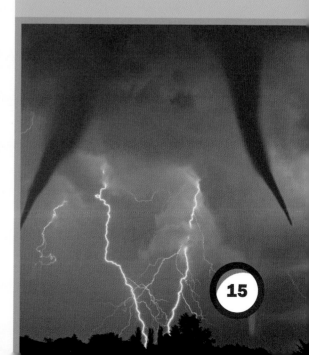

Warning the Public

Storm Prediction Center keeps a close eye on tornadoes.

Doppler radar also spots tornadoes. After an alert, the public must act fast. People must reach shelter.

Many communities use warning sirens. These sirens alert people to hurry inside. Weather radio stations issue storm updates. Smartphone apps can also track updates.

NOAA is the National Oceanic and Atmospheric

Tornadoes happen fast. It is important to be alert during a storm. The Storm Prediction Center (SPC) is in Norman, Oklahoma. It watches for tornadoes. It issues tornado alerts.

A tornado watch means twisters may form. A tornado warning means a tornado has been seen. People act as tornado spotters.

Administration. NOAA tracks weather data 24 hours a day. They share emergency information with news stations.

THINK ABOUT IT

The NOAA shares its data for free. Why do you think they do this?

750
Number of NOAA radio transmitters

- NOAA Weather Alert radios will set off a loud alarm if a tornado is coming. The alarm works even if the radio is turned off.
- The SPC also monitors severe winter weather and fire weather events.
- Better forecasting and warnings have reduced the number of tornado deaths.

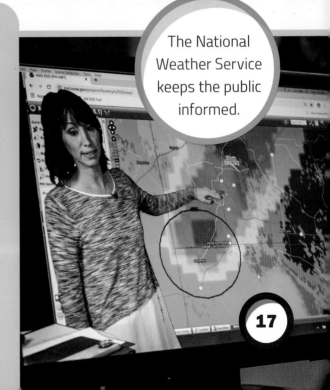

The National Weather Service keeps the public informed.

Shelter in a Storm

Tornado shelters protect people. Many storm shelters are built into the ground. In some places, the ground is too soggy. Underground shelters are not allowed in areas known to flood, either. In these places, storm shelters are built above ground. Sometimes a room in a house is strengthened to be a storm shelter.

Tornadoes toss a lot of debris around. Tornadoes can turn boards into spears. They can

punch through solid walls. This means a storm shelter must survive an impact from a heavy board moving at 100 mph (63 kph).

Some towns and cities provide community safe spaces. People far from home can shelter there. If a house is old or weak, people should use the community safe space.

Schools have severe-weather drills. In tornado areas, some schools have built tornado-safe rooms. These can often be a school gym.

250

Wind strength in miles per hour (400 km/h) that safe rooms must be able to survive

- Engineers test tornado shelter materials using a tool like a giant potato gun.
- Settlers on the prairies dug storm cellars.
- In some places, residents register their safe room so rescuers know to look there after a tornado.

19

A Trail of Destruction

They destroy trees and wildlife homes. Hailstones may kill animals. Animals may drown in floods. Lightning strikes may spark fires.

Usually, the impact is brief. Birds and animals may run away from an area. Later they can return. Damaged trees and uprooted plants will grow again. Nature is used to rebuilding.

Sometimes tornado damage is severe. Human garbage

Tornadoes usually move quickly through an area. However, all the wind and whirling junk cause damage. People, pets, and farm animals may be harmed.

Tornadoes also harm the natural environment.

can pollute water sources. Dangerous household chemicals also pollute areas after tornadoes. Fuel containers or pipelines can burst. These may release oil, sewage, and chemicals into the soil or water. These leaks can harm the environment for a long time.

THINK ABOUT IT

How might tornadoes help the natural environment?

30,000

Weight in pounds (13,608 kg) of a farm machine shoved in a 1995 Pampa, Texas, tornado

- Flash floods are common after tornadoes.
- A hailstone the size of a volleyball fell on South Dakota in 2010.
- Tornadoes can peel the pavement off roads.

After a Tornado

Rescue teams search for survivors.

People face many challenges after a tornado.

Sometimes people are trapped after a tornado. They need to let rescuers know their location. They might send a text message. However, cell towers and power lines may be damaged. They can

50
Percent of injuries that happen after a tornado, according to one study

- Stepping on nails is a common injury after tornadoes.
- Drones are used after tornados to find missing people.
- Search dogs are trained to find people who might be trapped.

bang on a pipe or blow a whistle. People should not shout for help. They may breathe in harmful dust.

Once it is safe, people should let friends and family know they are safe. Texts, the Red Cross "Safe and Well" site, and social media are good ways to do this. Unfortunately, the systems may not work until the power comes back on.

People are often evacuated from their homes. They are safer far away from damaged gas and power lines. Shattered buildings are dangerous. They can collapse suddenly. Emergency crews need time to clean up and make repairs.

EYES IN THE SKY HELP RESCUE EFFORTS

Drones are useful tools during disasters. Heat-detecting drones can locate victims trapped under debris. The drones can also send messages to victims waiting for rescuers.

The Fascination of Tornadoes

Storm chasers make exciting documentaries. These films allow people to safely experience a tornado's power. People do not have to get near the storm. They can watch on a screen.

Tornadoes appear in exciting stories, too. The famous book and movie *The Wizard of Oz* is about a tornado. It whirls a girl named Dorothy from Kansas to Oz. Oz is a magical place.

Fast and furious. Tornadoes can be scary. They are also fascinating. They inspire some people to become tornado scientists. Others are storm chasers. They learn about tornadoes and take pictures. Others like to watch videos or read about tornadoes.

A tornado is at the heart of *The Wizard of Oz.*

The book was published in 1900. That's 25 years before the deadliest twister in US history struck.

Since then, many movies include tornadoes. *Twister* is a popular tornado movie from 1996. It is about two storm chaser teams. They race across Oklahoma.

$500 million

Amount *Twister* earned in theaters

- In 1996, a tornado damaged a drive-in theater in Canada about to show the film *Twister.*
- FX team used a camel moan for the sound of tornado roars.
- They used a Boeing 707 jet engine for screaming wind.

Tornado Safety Myths

A dangerous myth tells people to open windows before a tornado strikes. People thought open windows relieved pressure in the house. That would stop tornadoes from smashing them. Open windows let in dangerous winds, flying junk and glass.

Should people take shelter under a bridge during a

Over the years, people have come up with ideas about tornadoes that are not true. These are called myths. Some myths are dangerous.

One myth says that greenish clouds always warn of tornadoes coming. Clouds turn green when they contain ice crystals. Also, some tornadoes arrive without a greenish sky.

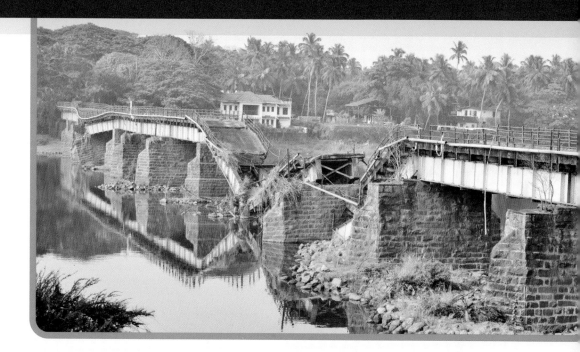

tornado? Definitely not! The bridge may collapse. Winds increase under bridges.

Another myth promises tornadoes never strike the same place twice. A church in Guy, Arkansas, has been hit by three different tornadoes. Oklahoma City has been hit over 100 times.

75

Age of a woman who flew in her bathtub in a 2017 Texas tornado

- Meteorologists do recommend sheltering in tubs because they are solid and anchored down.
- People once thought tornadoes turned green because they sucked up frogs or grasshoppers.
- Cordell, Kansas, suffered repeat tornadoes in 1916, 1917, and 1918.

Staying Safe in a Tornado

There are several ways to keep yourself and your family safe in case of a tornado.

- Have a plan of where to go and what to do if a tornado comes.

- Know the signs of a possible tornado.

- If you hear an outside alarm, head for shelter right away.

- Have an emergency kit with food, water, first-aid supplies, flashlight, batteries, etc.

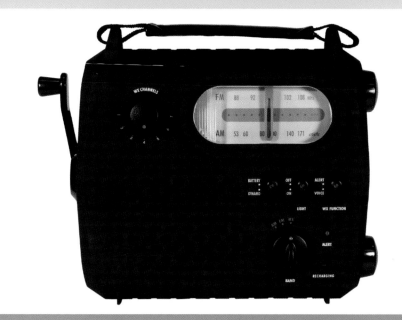

- Have an NOAA weather radio, and use it.

- If you are in a car, try to get to a building.

- If you are in a mobile home, try to get to a stronger building.

- Go to a storm shelter or basement.

- Wear strong shoes. Put on a bike helmet. Use your arms to protect your head and neck.

SEVERE WEATHER
SHELTER AREA

Glossary

counterclockwise
The direction opposite to the direction a clock's hands move.

debris
The broken pieces of structures, vehicles, and plants left after a destructive tornado.

Doppler radar
A system for detecting air motion in a storm.

ecosystem
Everything that exists in a particular environment.

evacuate
Leave a place during a time of danger.

flash flood
A flood that happens very quickly with little or no warning.

habitat
The natural place animals or plants usually live.

horizontally
In the same direction as level ground or the horizon.

outbreak
A group of tornadoes that happen around the same time.

radar
A system for detecting weather by sending out waves that reflect off precipitation.

waterspout
A mass of spinning air and mist that stretches from a cloud to a body of water.

whirlwind
A column of spinning air that forms a cylinder or funnel shape.

Read More

Baby Professor. *What to Do Before During and After a Tornado.* Newark, DE: Speedy Publishing, 2017.

Carson, Mary Kay. *Inside Tornadoes.* New York, NY: Sterling, 2010.

Challoner, Jack: *Hurricane & Tornado.* New York, NY: DK Children, 2014.

Raum, Elizabeth. *Surviving Tornadoes.* North Mankato, MN: Raintree, 2011.

Rudolph, Jessica. *Erased by a Tornado!* New York: Bearport Publishing, 2010.

Visit 12StoryLibrary.com

Scan the code or use your school's login at **12StoryLibrary.com** for recent updates about this topic and a full digital version of this book. Enjoy free access to:

- Digital ebook
- Breaking news updates
- Live content feeds
- Videos, interactive maps, and graphics
- Additional web resources

Note to educators: Visit 12StoryLibrary.com/register to sign up for free premium website access. Enjoy live content plus a full digital version of every 12-Story Library book you own for every student at your school.

Index

animals, 11, 20

Bangladesh, 14, 15
birds, 11, 20

cleanup after a tornado,
    22–23
climate change, 15
colors, 8
cone tornadoes, 8

damage, 15, 20–21
dangers, 18–19, 21
deaths, 15
Dixie Alley, 6–7
drones, 23
dust devils, 5

forecasts, 10–11, 12–13
formation of tornadoes,
    4–5

history, 14–15

injuries, 23

meteorologists, 10–11,
    12–13
movies, 24–25
myths, 26–27

National Oceanic
    and Atmospheric
    Administration (NOAA),
    16–17
National Weather Service
    (NWS), 13
number of tornadoes,
    7, 13

rope tornadoes, 8

safety, 18–19, 26–27,
    28–29
shelter, 18–19
speed of tornadoes, 5
storm chasers, 24
Storm Prediction Center
    (SPC), 16
stovepipe tornadoes, 8
Super Outbreak, 15

Tornado Alley, 6–7
Tri-State Tornado, 14, 15
types of tornadoes, 8–9

waterspouts, 8
weather alerts, 16–17
width of tornadoes, 8

About the Author

Kathleen Corrigan is fascinated by science and the world around her. She has studied and taught science. She loves to share her excitement with her students and the people who read her books.

READ MORE FROM 12-STORY LIBRARY

Every 12-Story Library Book is available in many formats. For more information, visit **12StoryLibrary.com**